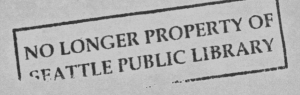

MAKING UP

YOUR

OWN MIND

MAKING UP YOUR OWN MIND

Thinking Effectively through Creative Puzzle-Solving

EDWARD B. BURGER

PRINCETON UNIVERSITY PRESS
PRINCETON AND OXFORD

Requests for permission to reproduce material from this work
should be sent to permissions@press.princeton.edu

Published by Princeton University Press
41 William Street, Princeton, New Jersey 08540
6 Oxford Street, Woodstock, Oxfordshire OX20 1TR

press.princeton.edu

Library of Congress Control Number: 2018954900
ISBN: 978-0-691-18278-0

British Library Cataloging-in-Publication Data is available

Editorial: Vickie Kearn, Susannah Shoemaker, Lauren Bucca
Production Editorial: Karen Carter
Text Design: Lorraine Doneker
Jacket/Cover Design: Karl Spurzem
Production: Erin Suydam
Publicity: Sara Henning-Stout
Copyeditor: Theresa Kornak

This book has been composed in Baskerville

Printed on acid-free paper. ∞

Printed in the United States of America

10 9 8 7 6 5 4 3 2 1

CONTENTS

8. The Future Has Arrived . . .

An invitation to apply effective thinking to the puzzles beyond this book

Further musings on a curriculum having no curriculum

THE JOY OF LEARNING

Creating a Meaningful Piece of the Human Mosaic

This book is written for . . .

. . . people who believe that education should be an uplifting, life-changing proposition designed to enable the individual to think more effectively, and create and connect ideas to make greater meaning through an ever-evolving mindset that allows one to persistently flourish.

. . . people who see themselves as students— whether their formal education is unfolding in front of them today or disappearing in life's rearview mirror—people who are committed to embracing empathy and open-minded doubt in the noble pursuit of improving their world by first improving themselves.

. . . people who enjoy practicing effective thinking through brainteasers that help generate new insights and original solutions to both the whimsical puzzles within these pages as well as the far more important puzzles within their own lives.

This book was inspired by . . .

. . . a half-century journey through formal education that began when I was a struggling young dyslexic student, then as a tutor, instructor, scholar, professor, author, online video math semicelebrity, and university president. Today I apply all these experiences as I continue to create educational growth opportunities with the hope of lifting the lives of others.

———————

Every aspect of our education should be viewed as offering a meaningful piece to the living mosaic that will inevitably define our self. The goal of that creative process is not the indoctrination of specific points of view, but rather becoming a truly independent and wise thinker and *making up your own mind*— that is, creating yourself.

This volume includes essays on this perspective of learning and living, on how academic institutions can change and challenge the educational status quo, and on how these philosophies and my colleagues have inspired an unusual course entitled *Effective Thinking through Creative Puzzle-Solving*. This book also allows the reader to experience personal growth by mindfully delving into the puzzles offered in that course. However, unlike traditional puzzle books, the ultimate goal here is not to solve the puzzles but rather to practice a way of thinking through the puzzles.

Engaged participation is essential in your learning—you must have agency over your own, ongoing education. Thus, to realize the greatest impact from this tiny book, you must intentionally use the puzzles ahead as a playground in which to practice elements of effective thinking that can then be applied to all of life's challenges and opportunities.

The two chapters following the puzzles offer thought-provoking prompts and deeper insights that arise from effective thinking. However, given the unusual expectation of active engagement, those two chapters are formatted so that reading them requires unusual effort. The first is printed upside down and the second is printed as a mirror image. Those chapters hold the promise to model effective thinking in action but will have maximum utility only after you have invested the requisite time and patience to first engage with the challenges by practicing the suggested mindsets.

As in my classes, I want you independently to generate your own "Ah-ha!" moments of excited discovery. So please do not be deterred by the inevitable frustration generated by the puzzles themselves or the puzzling format of the two epilogues . . . they were deliberately designed to provoke greater thought.

Hence, your mission—should you decide to accept it—is not to quickly solve these classic puzzles and move on, even if you have seen similar such puzzles before, but rather to explicitly practice the templates of thinking offered herein. So look upon these puzzles—previously seen or otherwise—with a fresh mind's eye and with a goal of applying the practices of effective thinking to see everything—the foreign as well as the familiar—in novel and richer ways. That is, thinking *through* these puzzles offers an opportunity for impactful learning and personal growth—the ideal and ultimate goals of a meaningful education. *Enjoy the mindful opportunities that await.*

MAKING UP

YOUR

OWN MIND

1

The Academy

Making Meaning through Impactful Education

Today we hear much talk about the economics of formal education (especially within higher education), in particular, the ever-mounting costs together with existential questions of dwindling *demand*. However, absent in the din of these debates is the basic issue of *supply*. What does formal education offer, and is that offering truly worth the investment of both time and money? In fact, it is interesting to observe that the phrase "formal education" is not even well defined.

Formal education has become a paradigm in which the goal is a diploma with (*idealis*) Latin words embossed into an elegantly encased rectangular parchment. That certificate and the degree it implies—high school, associate, baccalaureate, and other degrees—then lead to a starting job with a starting salary. That parchment is necessary in order to cash in on that initial income and thus has devolved into today's practical currency for formal education. Students demand that return on investment, and faculty, administrators, parents, and the curriculum oblige. Even the language that students and instructors speak reveals this basic axiom: *I have to get this requirement **out of the way**. I need to **get through** this course.* This language reveals a sadder truth: Formal education has become an obstacle course of mostly mindless hurdles and barriers we must traverse just to land that first job.

This reality has led many individuals, from out-of-the-box entrepreneurs to legislators to students themselves, to dream of shortening the lengthy ordeal required to check off all the boxes necessary to receive that diploma. If that credential is the ultimate goal, then I offer here an efficient solution: Print out a diploma at birth and award it to the newborn right there and then—mission accomplished.

A high-impact formal education, however, should be a truly transformational experience and thus cannot be hurried. Such a mindful educational journey requires time and reflection, especially in today's age of impatience, in which information and disinformation are constantly distracting us through our personal devices. In my mind, a diploma and starting job are not the goals of one's formal education. Instead, the intellectual journey toward personal growth is the genuine goal of education. With impactful formal education, that first job is not *the thing*; the journey itself is *the thing*.

Thus, that starting job becomes just one important outgrowth of that larger, life-enhancing experience.

Moreover, the richer and more meaningful that intellectual journey, the better that initial job and the greater the opportunities for future success. We are living in an age of obsolescence in which any fixed body of skills today will become dated and perhaps useless tomorrow. The best way to flourish in your professional life is to allow yourself to flourish as your authentic self—and that requires you to explore a diverse landscape of human thought, see that panorama holistically, and then discover your genuine intellectual passions and beliefs.

One's formal education, with the correct focus and commitment, can offer this meaningful exploration into the world of knowledge and ideas. That focus cannot only involve *thinking about* a subject (which stops at the topic's edge) but needs also to include a focus on *thinking through* that subject, that is, first learning and absorbing the pathways of thinking that grew out of that study and then intentionally practicing and applying those mindsets to other areas of life to bridge ideas through a truly interconnected course of study.

But it is not enough for us to be merely presented with these opportunities; we must actively pursue them ourselves—which is neither straightforward nor comfortable. High-impact learning is often uncomfortable learning. We cannot pave over the rocky road to deeper understanding so that the journey becomes smoother or shorter. A robust and meaningful journey cannot be a mindless checklist of things to do, cross off, and move on. Nor can that journey be a segregated and siloed collection of subjects, facts, figures, theories, algorithms, and methodologies that will, for the most part, be quickly forgotten.

The goal of a truly impactful education is to mess things up: to challenge the basic assumptions of how one views the world and one's self, and to emerge from that journey with deeper insights into both. We might arrive at our undergraduate education knowing with certainty that we want to become a lawyer. But we need to let the journey move us and discover where it leads—yes, our initial plans may be messed up, and we might become a mathematician, university president, or something else entirely. In fact, our brains are not fully physically developed as seventeen-year-old human beings, so it is problematic to be making long-term decisions for our future selves before we are even fully ourselves.

Throughout our lives, we must remain open to intellectual journeys carrying us in new and unexpected directions. The point of an impactful and meaningful education is to allow individuals to flourish, to continue to grow and thereby change. This personal change is not the disruptive type that reprograms one's DNA and makes one into someone else. Rather, it is incremental and subtle improvements that might even be difficult to measure in the short term.

Southwestern University is artfully bucking the tide of conventional and lowered expectations in higher education and has embraced this novel commitment to the vision of impactful and meaningful formal education. With an unprecedented commitment to inquiry-based, active discovery-learning, and experiential education, we continue to craft distinctive programming that focuses on the life of the mind.

In February 2017, the Southwestern University faculty unanimously passed a new curriculum that includes a commitment to offer, in every course, intentional opportunities for students not just to think *about* the material but also to think *through* the material. By thinking through a subject, students discover the utility and power of that thinking beyond the subject itself. These different templates of thought offer different lenses through which to see the world in a richer, sharper, and more interconnected way.

Now, in every class, students are challenged to make connections between seemingly disparate areas by applying the thinking from one area to amplify their thinking in another. Perhaps the mindset honed to truly see works of art in an art history course will allow a student to see details otherwise missed within the plasma membrane of a cell viewed through a microscope in a biology class. Perhaps the search for a pattern as practiced in a mathematics course will enable a student to see structure and nuance hidden within a poem studied in a literature class.

Creating connections requires practice, and the initial attempts are often modest and somewhat superficial. But first attempts should not be final attempts. Intellectual growth arises from practice and patience, and thus cannot be hurried. At Southwestern, we call this unique commitment to thinking beyond the course material and making intentional connections *Paideia*. The ancient Greek word *Paideia* (παιδεία) originally meant the education of the ideal member of society through what today we refer to as the *liberal arts and sciences*. At Southwestern University it is the name of the unprecedented commitment to think through the material within every course of study and connect that thinking with ideas and knowledge beyond the course itself. This practice of the mind offers lessons that last a lifetime. It is also one that can be adopted widely and applied at all levels of learning.

Today, on my campus, when students independently discover an otherwise unforeseen connection, they often exclaim, "I've just had a *Paideia moment!*"—an exciting moment of meaning and deeper understanding. They reveal for themselves that which would otherwise have been hidden. They make the invisible visible— which is at the heart of original thought and creativity.

After actively engaging with these ways of thinking, creating, and connecting through every course of undergraduate study, students emerge from their interconnected intellectual Southwestern experience prepared to make meaning and make a difference—both within their world as well as within themselves.

I invite you to engineer your own *Paideia moments* of discovery and always remember that with education and lifelong learning: *The journey is the thing.*

2

The Course

A Lesson Plan for Life

In the fall of 2015, I decided to follow Southwestern University's distinctive *Paideia* philosophy to its extreme conclusion. I created the course *Effective Thinking through Creative Puzzle-Solving*, although transcripts list it as *Effective Thinking through Creative Problem-Solving*, as employers value clever problem solvers and might not see their challenges as the puzzles they truly are. In reality, we all face puzzles every day in our lives—personal, professional, small, and existential ones. Some of them can be cast in the negative light of *problems*, but life's puzzles extend far beyond life's problems. The remainder of this book offers the opportunity for the reader to personally experience this course (with an expanded version of this course overview in the appendix).

Although this class is the most profound course I have ever taught, I playfully call it the "*Seinfeld* of the curriculum" because the course is about nothing yet attempts to teach everything. It contains no short-term content; instead there is only a long-term goal—offering the answer to what I call The Teacher's 20-Year Question: *Twenty years from today, what will my students still have with them from their experience in my class?* I hope that my students will joyfully practice mindsets that will lift their creativity, enhance their abilities to make connections, and strengthen their aptitudes to think effectively throughout their lives.

These mindsets are practiced through a sequence of puzzles, three each week: one relatively straightforward, one a bit more difficult, and one that is intended to be truly challenging. All, however, are designed to provoke thought. The ultimate goal is not to solve the riddle at hand, but rather to apply multiple practices of effective thinking to see that puzzle in as many different ways as possible.

Solving the puzzle is like receiving the diploma—it's **not** *the thing*. The mindful journey that led to an imaginative insight or solution is *the thing*. That journey will promote the intellectual agility to initially see features of our world one way and then apply practices of effective thinking to see those same features in a brighter light. However, moving from that natural temptation for a quick solution to an enlightened desire to embark patiently on a thoughtful journey and discover where that path leads is extremely challenging without considerable practice. These puzzles offer the practice needed to make that enlightened perspective less challenging and more natural.

And practice is the key word: As you will see in the next chapter, although many of the prompts offered to provoke effective thought initially appear straightforward, the challenge is to embrace them as one's own and incorporate them into one's patterns of creativity and everyday thought. I hope that as you move through this book and journey through your life, you too will commit to being open to new templates of thinking and modes of analysis. They hold the promise to carry you to even greater heights.

You are now invited to engage in this unique curriculum and enjoy the intellectual stimulation fostered in this class for yourself. You can explore different forms of mindfulness and practice effective thinking through the puzzles and prompts herein. Embracing the practices of an effective thinker requires individual agency and ownership. You cannot passively sit back in an entitled position and demand, "Educate me." Instead you must intentionally generate your own opportunities to challenge and, over time, transform yourself. Your schools, teachers, professors, mentors, and even this book can play at most supporting roles in the story of your intellectual journey—you remain the central character and hero in that thoughtful adventure that continues to unfold . . . that adventure known as life.

3

The Practices

Mindsets of Effective Thinking

Effective thinking is a fairly new phrase, as opposed to the more common *critical* thinking. Effective thinking includes the objective analysis that is typically associated with critical thinking, but also includes broader modes of creativity, originality, engagement, and empathy. Also, there is a judgmental—often negative—connotation to "critical" that is absent in "effective": Our world would be a better place if we fostered a society of thoughtful citizens who were truly *effective* rather than solely *critical.*

The templates of mind that foster effective thinking were formally introduced in the book *The 5 Elements of Effective Thinking*† and informally presented in various other works and outlets, including the weekly series *Higher Ed* produced by Austin's NPR affiliate KUT-FM (available through the Internet as a podcast). *The 5*

† E. B. Burger and M. Starbird, *The 5 Elements of Effective Thinking,* Princeton University Press, Princeton, NJ, 2012.

Elements of Effective Thinking remains the best reference for the nuanced details of these practices of thought and is a strongly suggested reading to complement this mindful treatise.

This chapter provides an overview of those five elements, some practical ways of applying them, and a real-life story to illustrate independent effective thinking in action. The puzzles ahead offer you *minds-on* practice in using these techniques of effective thinking. The hope is that by practicing this type of thinking through these little puzzles, you will naturally apply that same effective thinking to the larger puzzles in your everyday life.

The five elements of effective thinking are interconnected guideposts to active learning and ongoing growth. They begin and end with the synergistic notions of ***deep understanding*** and ***change***. To achieve these two goals, in practice, and create thought-provoking pathways forward, one must include ***effective failure***, the art of ***creating questions***, and an appreciation for ***the flow of ideas***. These interconnected mindsets form the foundation for *The 5 Elements of Effective Thinking*. For each puzzle that follows, your challenge is to apply each of these elements until some hidden opportunity, structure, or pattern is uncovered. At that moment, you understand the puzzle more deeply and, therefore, change how you see the challenge. Ideally, that path of thought will lead to a new insight that will guide you to a solution.

For each of the following actions to spark your own thinking, you must commit to that mindset and invest the requisite time to stimulate effective thought. Each of these elements will first appear as "obvious" and deceptively basic, but each offers a powerful way to provoke your thinking and sharpen your mind. In fact, none of these prompts are "easy" to apply; each requires practice—and that is the purpose of the puzzles. Reading and even rereading the paragraphs that follow without taking action will have only limited impact on how you think. You must try each and harness the patience to struggle—not with solving the puzzles, but mindfully applying the practices of effective thinking through these puzzles into your everyday life. So return to these prompts again and again and use them to provoke your thinking on all matters large and small, serious and frivolous.

Understand Deeply.

If you ask someone a "Do you understand?" question, you will most often receive one of two responses: "Yes" or "No," neither of which is correct. Understanding is a spectrum rather than a binary proposition—and wherever you are in your current understanding, you can, with intentionality, understand more deeply. This reality, which offers a fine working definition of *formal education*, is perhaps one of the greatest triumphs of the human spirit: Intellectually, we can intentionally always delve deeper. Taking the time just to embrace this mindset can be truly transformational.

No matter the circumstance, assuredly assert to yourself that you do not fully understand the issue at hand. That assertion will automatically place you in a different mindset. If you assume ignorance, you will be open to discovering gaps in your understanding. Motivate yourself by reacting to the declarative prompt, "There are aspects of this issue I don't understand. I now must uncover them and work toward making greater meaning." Intentionally understanding more deeply, and thus seeing an issue in a different way, is a daunting challenge. Here are three practical ways to tickle the mind and provoke effective thought.

Start with the simple.

Understanding simple things in unusual depth is an oft-overlooked but powerful way to see greater nuance within the complex. When facing a serious challenge, start with a basic or even trivial version of it, in which you have a firm intellectual foothold. Now probe that simple scenario more deeply to see the detail and structure that always lies beneath the surface. Once you expose that complexity in the simple, you will see the original challenging case with greater clarity. But this suggestion is a challenge. It is difficult to stop and focus on that which you already believe is obvious and invest time to see that trivial state in a different way. This prompt generates incremental progress—by starting with the basics and practicing the patience required to probe that easy circumstance with exceptional depth, you take an important, mindful step toward understanding the original, multifaceted issue more deeply. Start with the simple and use it to divide and conquer the complex.

Spotlight the specific.

Often warming up with a special case or specific example is a strategic way to gain some new insight that can then be extended to the general situation. Look at an issue from the micro-level with the goal of seeing some hidden structure or pattern that persists at the macro-level. When considering a special case, reframe any particular structure discovered in that example to expose some general principle hidden in the original issue. It is only within that recasting that important insights are created.

Add the adjective.

To understand anything in greater detail, challenge yourself to add as many descriptors as possible, and take the time to consider each adjective and draw some new insight into the issue at hand. Do not leave an adjective for another descriptor until some new facet is revealed. By adding an ever-growing list of descriptors, you can uncover hidden confusion or misunderstanding, and, if the situation involves multiple perspectives, you can also realize greater empathy.

Fail Effectively.

Despite societal pressures and short-sighted norms, failing effectively is one of the most important pathways to deeper understanding as well as discovering new knowledge. Jumping over the cultural hurdle that "failing is bad" empowers you with an easy step forward. You might not *always* know how to do something correctly, but you can certainly *always* do it wrong; and in doing so, you can then *fail effectively*: Focus on that failed attempt and use that small misstep as a giant leap toward a deeper understanding and an inevitable resolution of the original issue.

Again, observe the incremental nature of this process. You are not trying to conquer the entire mountain; instead you are just thoughtfully taking one step and seeing what can be learned from there. Many find the celebration of effective failure vexing, but to be clear: Failing is not the final destination. Rather, effective failing is an important—often requisite—intermediate (mis-)step. If you consider yourself on a chessboard, then failure is a square you land upon with the sole purpose of moving off in a direction that would not have been possible without that intermediate move. However, failure becomes effective only when that mistake or failed attempt is not dismissed until you gain some new insight into the issue at hand. There is no greater teacher than one's own mistakes. But to learn effectively from that brilliant teacher, one must doggedly stay with that failed attempt until a new lesson is realized.

Fail fast.

Free yourself from a focus on perfection and instead focus on process. Thus, fail quickly and cheerfully: Whatever the task at hand, try doing it quickly and lousily. If authoring a document, do not stare at a blank screen; instead, write a miserable draft by letting your stream of consciousness flow. Now, instead of a blank screen, you have something to which to respond: your first, crummy draft. Respond to it: Find the hidden gems as well as uncover ambiguity and lack of clarity in your own mind about the issue. Revision and editing are writers' ways of effectively responding to earlier, requisite failed drafts. So get that first failed effort out of the way as quickly as possible and then start revising, rethinking, and learning from what you first created. Again, an epiphany needs to arise from that initial effort for it to be truly effective.

Fail again.

The comedian Steven Wright once said, "If at first you don't succeed, then skydiving definitely isn't for you." How true; but not succeeding offers a brilliant parachute that safely carries you to new discoveries. Suppose someone gives us a great challenge and we go off and try to resolve it and, it turns out, we fail. Ordinarily we feel discouraged. However, suppose instead that as this someone presents us with that great challenge, she also informs us that to realize success at this enormous task it is required that we first fail ten times. With this new information, our mindset changes when we make that initial failed attempt. We now think, "One down, nine to go; we've made *progress.*" But, as always, the progress comes only *after* the error: when you take the time to analyze that failed effort and let it carry you to a new insight. So, embrace this need to make ten initial mistakes. With ten failed attempts at your disposal, be open to being wrong and doubt those aspects of which you are certain—see where you are led and what, if anything, breaks down.

Fail intentionally.

Every failed attempt is an invitation to understand the situation more deeply by exploring why that effort did not work. Following this line of reasoning to its logical extreme, if you want to understand more deeply, you should *intentionally* fail to generate that epiphany or new perspective. Therefore, consider extreme cases and remove all real constraints to create completely impractical thoughts and solutions. With those wild ideas in mind, now see how they can be tamed or shaped into clever practical solutions that would never have been found without that initial, impractical, failed attempt. Determine the precise breakpoint where things went wrong. Study that breakpoint as well as what around it has promise—they might lead to a novel insight. More generally, start with any wrong approach or answer and then force yourself to delve into that error until you see some aspect of the original challenge amplified in a new light.

Create Questions.

The most straightforward way to probe more deeply is to create questions. Creating questions—even if those questions are not asked—moves us from being a passive bystander to an active participant in our life's journey.

Embracing a dynamic mindset of always generating questions will lead to deeper understanding. And you should demand this point of view from yourself as well as from those with whom you surround yourself. Never ask a group, "*Are* there any questions?" for you should have the expectation that everyone is truly engaged. Instead, offer the prompt, "*What* are your questions?" Or, "*What* are your questions that you wish to share with the group?" Prompting yourself in this same way will not only amplify your innate curiosity, but will also lead you to new discoveries.

Be your own Socrates.

Asking meta-questions throughout any thoughtful process will always shine a light onto the big picture and often force you to focus on the right challenge. Asking, "What is the real issue here?" opens your mind to the possibility that you are considering the wrong question or problem. For example, instead of trying to fix the frustrating traffic congestion on your commute to work, you might wonder how you could make that lengthy travel time less frustrating or more productive. Asking, "What if . . . ?" can refocus your thinking as you consider alternatives. Be open minded and ask big questions to discover the big picture.

Create basic questions.

Ask fundamental questions to make fundamental breakthroughs. Even wondering, "What does the simplest case look like?" and "What happens in that trivial situation?" are powerful ways of probing into the original, subtler scenario.

Ask something else.

Whether you are stuck or not, considering something else not only resets your thinking, but also allows you to refocus on the issue in an entirely original way. Asking, "What's a different but related question?" or "What's the opposite point of view?" will allow you to consider the issue from a diversity of perspectives and can generate a diversity of new insights and ideas.

Go with the Flow of Ideas.

When someone has a new idea, it is often cause for celebration: "Bob just had an idea—let's get balloons and a cake!" Although any excuse to celebrate with cake should be welcomed, in truth, the birth of an idea is always a beginning and never an ending. It is only after a new insight or idea is realized that the real creative heavy lifting begins by asking: What comes next? By considering how to connect a new idea to something else, by generalizing it to a larger context or by applying it to an unrelated situation, we are engineering our own creativity and not only provoking thought, but also provoking innovation. However, taking a cutting-edge idea and imagining what comes next is never easy.

When I was younger, telephones were boxes affixed to the wall with rotary dials and were rented from *Ma Bell*. Then came touch-tone phones, then phones you could own, then brick-like cellular phones. Today, smart phones can connect to your watch, make calls, show videos, take photos, and help you make instant restaurant reservations. As a child, I wondered if the science fiction flip-phone-looking communicators of *Star Trek* would ever become reality. Today, as I have witnessed this flow of ideas and technology take us to an age in which we can call each other on our watches, it is now difficult for me to imagine what comes next.

For today's children, however, the only phones they know are "smart"—that is where they start as they wonder if some science fiction fantasy today can become the reality of tomorrow. After the journey to create (or even observe) a new idea, it is a great challenge to forget the exhaustion of that long intellectual journey, and just look at where you are as the launch point for what is to come next.

Challenge yourself to let every new idea you encounter inspire the fresh, childlike response: "What is next?

How can I repurpose, extend, or otherwise generalize or reapply this new notion?" Go with the flow.

Run down all paths.

Whenever you are able, consider all possible cases, even the obviously impossible ones. Follow the flow of each scenario to its very end. Most, if not all but one, will lead to dead ends. But then learn from those failed attempts and apply that new knowledge as you travel down yet another possibility to its ultimate conclusion. Whenever there are only a few potential outcomes, considering them all and discovering why most cannot happen will allow you to discover what must, in fact, unfold.

Embrace doubt.

Challenge your own narrow thinking and opinions to see where that flow takes you. Embracing a diversity of points of view empowers you to see the multifaceted nature of complex things. In fact, empathy often allows you to see a situation in a totally different light and offers a perspective you would have never seen otherwise. Thus, consider the opposite side or alternative situation; contemplate the counterintuitive. Look at an issue from all angles. If it is a political or societal issue, remember that empathy and sympathy are not the same. You can empathize with another point of view without sympathizing with that side. Embrace doubt as a strength. Wonder, "What if I'm wrong?" and let your mind flow over the reasoned consequences of that supposition and see where you are led. Remember that the opposite of doubt is not certainty, but rather closed-mindedness. Always be open minded.

Never stop.

As with all of these elements of effective thinking, following the flow of an idea requires persistence and tenacity to see where that flow will carry you. Do not let go of an idea until it takes you somewhere new, unexpected, or to an insight into something otherwise unrelated. Every new idea is a beginning, not an ending. Thus, never stop the flow of your ideas.

Be Open to Change.

Applying these elements of effective thinking through these suggested prompts is not easy. Only over time will these habits of mind become practices of mind. That transition captures the quintessential element of education as well as thinking: that of change. We look at the world of ideas, nature, each other, and ourselves differently when we look at them through the different lens of effective thinking. Meaningful education is built on the reality that we are truly capable of change—not the change that reprograms us into people we are not, but that rather, over time, steadily makes us into better versions of ourselves.

Change is scary and sometimes threatens us on an existential level. The change encouraged here is ongoing, gradual, and evolutionary rather than sudden or disruptive. Small and incremental changes will transform how we think and engage with the world, but that journey unfolds over time—and although we cannot rush it, we can foster it. In fact, a healthy and natural state of being is one that is in continual flux. We should be ever changing, and our education should promote mindsets that support this dynamic perspective—one that encourages wise and creative thought with the openness to learn, grow, and change—so that deeper understanding is realized and new discoveries are made.

Thus, as you mindfully engage with the puzzles ahead, be conscious of how, through effective thinking practices, the puzzles themselves change: The way you first saw them in your mind will be different from how you will see them after you have challenged yourself to understand them more deeply. The ultimate goal is to change how you think about puzzles, not only those from this book, but those arising throughout your life.

These prompts to provoke thought are abstract until there is a concrete context in which to apply them for yourself. The puzzles ahead provide such a challenging context to practice these mindsets. To illustrate these elements of effective thinking in action and how people of all ages possess the innate ability to provoke their own effective thought, I close this chapter with an aspirational true story.

Seamus Succeeds.

Late one July, while visiting friends in Massachusetts, I was asked by their son Seamus, who had just completed third grade, if I might help him with some unpleasant summer mathematics homework. One question caught my eye:

> You have 36 doughnuts and you wish to arrange them into two rows each having the same number of doughnuts. How many doughnuts would be in each of the rows?

To you and to me this question is asking, "What is half of 36?"; however, to a third-grade Seamus it was asking, "What number plus itself equals 36?" Seamus was a very bright young student, and I asked him if he understood the question, to which he replied, "Yes," and proceeded to pick up a pencil and put it to a sheet of paper. Time passed and nothing happened—he was frozen. His brow was furrowed as if he was trying to "think harder." He was, in some sense, intellectually constipated: pushing and pushing with no idea coming out.

As these elements of effective thinking demonstrate, thinking better does not necessarily require us to think harder—instead, we have to think *differently*. So I asked Seamus to stop doing whatever it was he was trying to do, and his face was instantly transformed from that troubled state to its usual cheerful countenance. I then directed him with, "Seamus, when I say 'GO', I want you to quickly give an answer you are confident is *wrong*." He looked at me quizzically. I asked, "Are you ready?" to which he responded, "I guess . . ." I said, "GO!" and he immediately shot back, "16!"

Certainly, 16 is a wonderful wrong guess. The number 36 ends with a 6, as does 16—a nice digit pattern; also, he was searching for half of 36 and 16 is, at least, smaller than 36 (had he answered, say, "116," I would have worried)—so he had made incremental progress. I did not share these spontaneous thoughts of mine with Seamus. Instead, my very next words to him were, "Great. Now show me why your answer is wrong." Seamus now had something concrete to do: He could react to his guess. He carefully and correctly added 16 to 16 and discovered the sum to be 32. He looked at that answer for a moment and then exclaimed, "Oh, that's too small . . . The answer must be 18!" And so it was.

Notice that when Seamus was focused on perfection, he was almost catatonically frozen. However, when he was given permission to focus on the process and also to fail, he quickly and triumphantly came upon the correct answer. It is also revealing to review my role in this wonderful true story. I offered the following two prompts:

> "Quickly give an answer you are confident is wrong."
>
> "Great. Now show me why your answer is wrong."

I did not teach him any mathematics; in fact, he did not even need me: Had he challenged himself with those two generic prompts, to *ask* for a wrong answer, effectively *fail*, and follow the *flow of that stream of thought*, he would have guided himself to an epiphany and subsequently a *deeper understanding* as well as a correct answer completely on his own. Thus, he would have moved from a confused and frozen state to an enlightened and successful one; that is, he had it within himself to *change*. And so do you.

4

The Future

Unleashing Your Inevitable Genius

Throughout higher education we find examples of educators who want their students to fix an endless list of problems and injustices within the world. They do so by painting a bleak picture of reality that depresses the soul and exhausts the mind. Those faculty members are often frustrated, and their students quickly mirror those emotions and soon become fatigued, upset, overwhelmed, and ineffective. Those educators are—with the best of intentions—trying to save the world by indoctrinating their students to see only the problems, and thus training them to be problem focused.

> I arise in the morning torn between a desire to improve the world and a desire to enjoy the world. This makes it hard to plan the day.
>
> —*E. B. White*

Everyone deserves to be inspired to be opportunity focused and effective in realizing any goal they set for themselves—including finding novel solutions to the many real problems of our time. To that effective end, it is better first to uplift individuals through a challenging, joyful intellectual experience that then empowers them to uplift the world. We make the world better by first making ourselves better.

However, sometimes we are not the best judges of ourselves—we are often far too modest and thus we do not always challenge ourselves to realize even greater expectations. We are all mosaics made up of those who have inspired and taught us: family members, friends, teachers, mentors, and even strangers. We continue to assemble those individual pieces as we make up our minds and create ourselves. Such transformational change is difficult because it is impossible to see another potential self or future reality. And that ambiguity is precisely the challenge: to move forward gradually and incrementally without perfect information as to where your effective thinking will lead you. That winding but thoughtful path makes a difference—emotionally, physiologically, creatively, and intellectually.

I hope you enjoy the window into my *Seinfeld*-esque class opened through this book and that the puzzles ahead will be an enticing invitation for you to search for opportunities to provoke your own thought. Create the time and the space for mindful solitude. Carve out moments of stillness—away from both your living and electronic companions—in which you can quietly reflect on your thoughts and emotions in the present.

Practice the contemplative art of effective thinking. Through those intentional moments of mindful peace, you will not only reawaken deeper empathy that is within us all, but you will also rediscover greater joy—joy that is all of ours to hold.

This mindset does require a leap of intellectual faith—faith in your own innate abilities to think, create, and connect—and a firm belief that you can take practical steps to be even better than you currently are. That positive future offers one of the pleasures of living: to be an ever-evolving, ever-growing, ever-learning, and ever-improving being who is opportunity focused and is always searching for new and brighter sunrises while appreciating all the beautiful and inspiring sunsets offering the promise of another tomorrow.

5

The Puzzles

Much Ado About Nothing

For each puzzle, your challenge is to apply all five practices of effective thinking and see the puzzle from as many different perspectives as possible. Feel free to revisit Chapter 3 again and again to help move your thinking forward. Success here is not in the solving of these puzzles, but rather in effectively thinking through them and then applying those same templates of thought elsewhere. Even if you have seen a puzzle before or remember its solution, take the time to apply effective thinking and challenge yourself to see it in a new way and, perhaps, create a new solution.

For any puzzle, if you would appreciate seeing that thinking unfold a bit, you are invited not only to turn over these puzzles in your mind, but also turn over this book and visit Chapter 6. There you will discover some specific thought-provoking steps for each puzzle. Chapter 7 then reflects further on where effective thinking can guide you for the puzzle at hand and beyond. However, for maximum impact, you should visit the reflections of Chapter 7 only after you have invested considerable time and effort for your own reflections, whether or not you resolved the puzzle. After you solve a puzzle, you are encouraged to read the associated commentaries in Chapters 6 and 7 to compare your process with possibly another perspective.

Remember that frustration is a healthy and often necessary part of true learning and growth. Thus, I hope you will see the reading (decoding?) of Chapter 7 as yet another puzzle to solve and also a metaphor for seeing how issues (and even words) can look very different when viewed from a different perspective. But even if you decide to pass on this invitation to engage with the more unusual features of this book, I encourage you to end your journey here by reading the concluding thoughts in Chapter 8: *The Future Has Arrived.*

For those readers who have focused only on getting an "A" in the class, I remind them that the solution here, as the diploma in formal education, is not *the thing. The thing* is the journey itself—the *process* of creation. Just as musicians develop their skills by practicing scales until their instruments truly sing, I encourage you to practice effective thinking through these whimsical puzzles until you can apply those practices naturally to the larger puzzles of life. So do not quickly solve a puzzle and move on, but instead take the time to think carefully through each and allow that reflection to lead you to your own *Paideia moments* of meaningful epiphanies. Effectively thinking through these puzzles will allow your mind to sing as well as to soar.

Finally, once you have thought through a puzzle, before turning your mind to the next one, invest the time to consider: *What effective thinking practices did I apply? What new insight or thought was provoked? How do I now see the puzzle in a different way?* Then challenge yourself to apply those lessons and practices in yet another context.

Enjoy the challenging pleasures of thinking effectively.

CHALLENGES I
Practice Effective Thinking through . . .

I.1. Who's Who?

One afternoon, on a college campus, two students—one a math major, the other a philosophy major—were conversing.

"I'm a math major," said the one with black hair.

"I'm a philosophy major," said the one with red hair.

Given that at least one of these students is lying, what color hair does the math major have?

I.2. When Six Equals Eight.

Draw six straight-line segments of equal length so that they produce eight equilateral triangles (recall that a triangle is equilateral if all three sides have the same length, which also implies that each of its angles measures 60 degrees). *Note:* There are many ways to answer this puzzle—use different practices of effective thinking to create as many different solutions as you can.

I.3. Cutting Boards.

Suppose you are given a standard 8 × 8 chessboard and a large supply of dominoes. Each domino can cover exactly two squares on the chessboard (see the leftmost chessboard on the next page). As a warm-up, verify that the standard chessboard can be covered completely by dominoes so that each domino covers exactly two squares and the dominoes do not overlap one another. Now suppose that the two squares of the chessboard have been cut off as shown in the middle image below. Your first challenge is to determine if you can cover this cut chessboard with nonoverlapping dominoes so that, again, each domino covers exactly two squares. Finally, your last challenge is to consider the same question for the truncated chessboard on the right. Justify your responses.

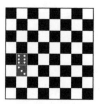

One domino
correctly placed
on the chessboard

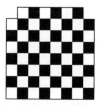

The chessboard
with two upper
squares removed

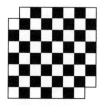

The chessboard
with two other
squares removed

CHALLENGES II

Practice Effective Thinking through . . .

II.1. Three Switches, Two Rooms, and One Bulb.

Two windowless rooms in an unusual building are connected by a long and winding hallway—so winding that it is impossible to see any part of one room while in the other. The first room has three identical light switches on the wall, all in the down (off) position—two of which do nothing and the third of which turns on and off an old-fashioned desk lamp that sits on a table in the second room. What are the fewest number of trips up and down that winding hall required to determine which switch is the one that controls the lamp in the other room?

II.2. Going from 5 to 4 in Two Moves.

In the figure below, you see five 1 × 1 match squares. By just changing the positions of two matches (you cannot break or remove any match), change the number of 1 × 1 match

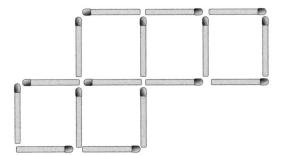

squares from five to four. Note that each match must be a full side of a match square—that is, no loose ends or dangling matches are allowed. Furthermore, you are not allowed to place one match on top of another.

II.3. A Slow Burn.

You are given two pieces of string—not necessarily of the same length—and are told that each piece takes exactly one hour to burn completely when an end is lit on fire, but that the burning is **not** uniform (that is, it will not necessarily take 30 minutes to burn half of the length of either string). With just these two pieces of string and a few matches, but without any other timepieces whatsoever, is it possible to time exactly 45 minutes? Explain your response.

CHALLENGES III
Practice Effective Thinking through . . .

III.1. A Top 10 List.

For each of the following ten statements, determine if the statement is true or false:

1. Exactly one statement on this list is false.
2. Exactly two statements on this list are false.
3. Exactly three statements on this list are false.
4. Exactly four statements on this list are false.
5. Exactly five statements on this list are false.
6. Exactly six statements on this list are false.
7. Exactly seven statements on this list are false.
8. Exactly eight statements on this list are false.
9. Exactly nine statements on this list are false.
10. Exactly ten statements on this list are false.

III.2. Five Elements, but Only Four Hats.

Four bright students focusing on effective thinking (**A**licia, **B**ryce, **C**arol, and **D**evin) volunteer to take part in a thought-provoking puzzle. They each agree to be buried up to their chins in a large "ball pit" (a pool filled with colorful hollow plastic balls). They are arranged in a straight line and agree to be very still—so they can only look straight ahead. There is a large, opaque, nonreflective screen with the motivational words **"KEEP THINKING EFFECTIVELY"** written on both sides and separating one of the students from the rest (see the figure below). The students are lined up so all are facing the screen for inspiration—which they cannot see through—and know where each classmate is standing. The person hosting this puzzle then places a hat on each of the four students exactly as shown in the figure. There are two black and two gold hats, and the students know this fact but do not know their individual hat color. Each of the four of them will be awarded a $100 prize if any one of them correctly calls out the color of the hat on her or his head; otherwise, each student's tuition bill will be increased $1000 as a penalty. The four students are not allowed to talk to each other, and have 10 minutes to figure something out. Recall that **A** and **B** can only see the **"KEEP THINKING EFFECTIVELY"** sign, **C** can only see **B**, and **D** can see **B** and **C**. After one minute, one of them calls out with the color of her or his hat. Which student was the one to call out? Why is that person completely confident in being correct?

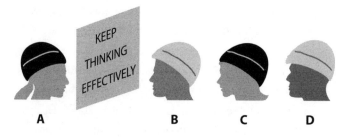

III.3. Roommates from Ruter.

"Hi! My name is Ralph and I live in the all-male luxurious dorm known as Ruter Hall with my roommate Rich. One

Saturday, we decided to host a ruckus Roto Ruter party in our room and we invited four other pairs of roommates from our floor. When everyone arrived, various handshakes took place. Obviously, no one shook hands with himself, and perhaps less obviously, no one shook hands with his own roommate. Also, no one shook hands with the same person more than once. After the handshaking stopped, I asked each of the nine other guys in the room how many hands he had shaken. Much to my amazement, each person gave a different number as his answer!" Here is your challenge: How many hands did Rich shake?

CHALLENGES IV

Practice Effective Thinking through . . .

IV.1. A Dismal Downpour.

This past spring, gallons of rain fell upon an already-green university, which made the main Academic Mall even more lush and beautiful than normal. On one particular soggy spring day, it began thundering, lightning, and raining cats and dogs at the stroke of midnight. Given the stormy weather patterns of that spring, is it possible that 72 hours after this thunderstorm began, the weather on campus was clear and sunny?

IV.2. Making 4 from 12.

A *polygon* is any geometric shape drawn with straight edges so that the edges do not cross each other; the end of *each* edge touches *exactly* one end of another edge; and the shape has both one inside and one, separate outside. There are many polygons that can be created using 12 matches. Here are two such examples:

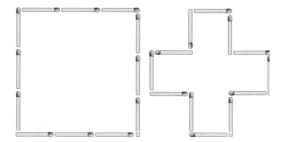

Notice that each of these polygons has a perimeter equal to 12 matches, and that the area of the first polygon (which is just a square) equals 9 (matches²), while the second polygon has an area of 5 (matches²). Can you use these matches to create a polygon having a perimeter equal to 12 matches and with an area equal to only 4 (matches²)?

IV.3. Map Folding.

Long, long ago, before the age of GPS and Google Maps, people used paper maps. There were two main defects with those old-time charts: First, there was no calming voice that offered audible directions ("recalculating . . ."); and second, you had to fold up the map when you were done using it—not always an easy task, if you wanted to fold it back in the same way in which it was originally folded. Consider the campus map on the next page that has numbers written on only the front side of the map. You are to fold the map just along the straight lines given (that is, folding only along the edges of the squares that surround the numbers) so that the "1" square is face up on the top and all the other squares are directly underneath that top square (so it will be folded to have the dimensions of a single square). Your challenge: Fold the map so that the numbers appear in *numerical order* from 1 to 8 (so the "1" square touches the "2" square, which is directly above the "3" square, etc., with the "1" square face-up on the top). Of course, you cannot cut the map in any way.

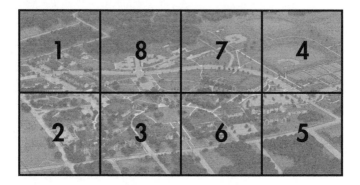

Bonus

Once you succeed in folding the campus map above, try the more challenging map folding sequence given below: Follow the same directions as previously given—namely, the map is to be folded only along the horizontal and vertical lines into a stack of eight squares in numerical order with the top face-up square showing "1." Again, as much as you might want to, you are not allowed to cut the map!

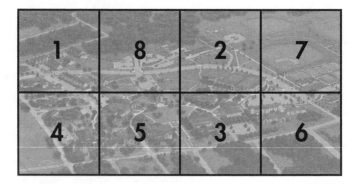

CHALLENGES V

Practice Effective Thinking through . . .

V.1. Whoops.

Among the assertions made in this puzzle, there are four errors. Can you find them?

- $2 + 2 = 4$
- $4 \div \frac{1}{2} = 2$
- $3 \frac{1}{5} \times 3 \frac{1}{8} = 10$
- $7 - (-4) = 11$
- $-10(6-6) = -10$

V.2. A Not-So Weighty Challenge.

You are presented with nine identical-looking stones and are told that a valuable gem discovered by a thoughtful pirate is embedded within one of these stones. The bejeweled pirate stone is imperceptibly (to you) heavier than the remaining eight stones (all of which are of identical weight). You are also given two extraordinarily inexpensive balance scales, each of which can be used only once before it breaks and cannot be used again. Is it possible, with just two weighings on the balance scales, to determine which stone holds the secret of the pirate's treasure?

V.3. A Star Is Born.

The standard five-pointed star drawn with five straight lines has five disjoint triangles (that is, no triangle is contained inside of or overlaps another triangle). Draw two straight lines that cross one five-pointed star in such a manner that the resulting drawing contains ten disjoint triangles. The stars here can be used for your first two attempts.

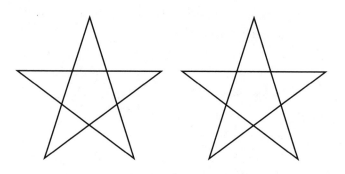

CHALLENGES VI

Practice Effective Thinking through . . .

VI.1. Puzzling Politicians.

One hundred politicians were gathered at a certain assembly to argue, debate, shout, and otherwise caucus. Each one was either honest or crooked, and we are presented with two additional facts:

> Fact 1: *At least one of the politicians was honest.*
>
> Fact 2: *Given any two of these politicians, at least one was crooked.*

Given all this information, can it be determined how many of the politicians were crooked? If so, what is that number? If not, why can we not know for certain?

VI.2. Turning on a Dime.

In the figure to the right, you see four matches arranged to form a "cocktail glass" and *inside* that glass is a dime. Your challenge is to move only two matches to rebuild the cocktail glass configuration of matches so that the dime is no longer *inside* the glass. Of course, you are not allowed to touch or move the dime. Reconstructing the cocktail glass so that it is upside down but with the dime still inside the glass does not resolve this puzzle; that is,

you cannot be sneaky and think that the dime will, on its own, fall out. The dime will not be moved by gravitational wishes.

VI.3. Penny for Your Thoughts.

A number of pennies are spread out on a table. They lie either heads-up or tails-up (see the sample illustration below). Unfortunately, you are blindfolded and so the coins are hidden from view. You can feel your way across the table and therefore can count the total number of pennies on the top of the table, but you cannot determine if any individual penny rests heads up or down (perhaps you're wearing puffy gloves). You

are informed of one fact (beyond the total number of pennies on the table, which you can determine for yourself): Someone tells you the number of pennies that are lying heads-up. Your challenge is, while remaining blindfolded, to move the coins around in any way you wish, or turn any of them over (as long as the final configuration has all the pennies resting either heads or tails up on the table), so that when you are finished, you will have divided the pennies into two separate collections on the table with the property that one collection has the same number of heads-up pennies as the other collection.

CHALLENGES VII

Practice Effective Thinking through . . .

VII.1. A Doggone Puzzle with Cats.

Ten pets are to be fed exactly fifty-six biscuits in total. Each animal is either a dog or a cat. Each cat is to be given five biscuits and each dog is to be given six. How many dogs must there be? *Bonus*: Solve this puzzle just using practices of effective thinking rather than resorting to doggone algebra.

VII.2. A Cross Farmer Needs to Cross a River.

Farmer Francis needs to transport a rabbit, a fox, and a bunch of carrots from one side of a river to the opposite bank. Our farmer has a tiny raft that can hold Francis and just one other passenger (either the rabbit, the fox, or the bunch of carrots). The problem is that if left alone without Francis, the rabbit would eat the carrots and the fox would eat the rabbit. Assuming that neither the fox nor the rabbit would run away if left alone on either side of the river, is it possible for the farmer to transport all three across the river with all three alive and uneaten?

VII.3. Making 50/50 Greater than 50/50.

You are presented with two identical bowls and 100 marbles—50 black and 50 gold—all of the same size and weight. You are invited to place the 100 marbles into the two bowls in any way or configuration you wish, as long as each marble is in one of the two bowls. Each bowl is shaken so the marbles in the bowls are thoroughly mixed. You are then blindfolded and the two bowls are placed in front of you in a random order. While donning the blindfold, you are to select a bowl and then reach into that bowl and remove just one marble. If it is black, you win; if it is gold, you lose. Now knowing all the facets of this game, how would you place the 100 marbles into the two bowls so the likelihood of winning would be greater than 50%?

CHALLENGES VIII
Practice Effective Thinking through . . .

VIII.1. The Deficient Digit.

Recall that a digit is any of the following numbers: 0, 1, 2, 3, 4, 5, 6, 7, 8, and 9. When juxtaposed, they form the numerical building blocks of the counting numbers, for example: 16, 28, 663. Consider all the numbers from 1 to 1000. Within those thousand numbers, which digit appears the least frequently and why? *Bonus*: Which digit appears the most frequently?

VIII.2. Going in Circles.

U-Turn University, with their slogan *At U-Turn U, We're Driven to Turn U Around,* welcomes everyone "thru" their main gate tollbooth and are equally proud of their students who own cars as well as those who do not. It was life in the fast lane for all until Provost Pothole hit upon a peculiar numerical speed bump as he was computing and comparing the grade point averages (GPAs) of the current graduating students to the graduates of the previous year—each graduating class having 1000 students. He discovered that the average GPA of all current graduating students owning cars is greater than the average GPA of all students owning cars who graduated last year; and that the average GPA of all current graduating students not owning cars is also greater than the average GPA of all students not owning cars who graduated last year. It looked as if the current graduating students were as bright as a halogen headlight. The problem is that the average GPA of this year's entire graduating class is, in fact, *lower* than the average GPA of last year's entire graduating class. Is this scenario actually possible, or must it be the case that Provost Pothole made an error and landed in an arithmetical ditch? *Note:* By "average" we mean "mean."

VIII.3. Slicing a Square in Half in Three.

Suppose you have eleven wooden matches, all of the same length. Using eight matches, you form a square with each of its sides equal to two match lengths. Is it possible to place the remaining three matches, touching end to end and with the two outer ends touching the square so that they divide the area of the 2×2 square into two equal areas? *Bonus:* If it is possible, can you find a way of cutting the area of the square in half so that the two ends of the chain of three matches touch the square at its corners? Or, can you explain why such a division is impossible?

AN EXCEPTIONALLY CHALLENGING BONUS
Practice Effective Thinking through . . .

The **C**razy and **E**xceptionally **O**bnoxious **CEO**.

You have a job that you truly love—you enjoy your work, your colleagues, and your handsome salary. One day, however, the CEO of the company calls you and everyone from your unit to a meeting. There he reports that there will be serious changes made tomorrow and goes on to explain what will happen. He will line up you and your colleagues from your area—one behind the other—in a straight line. He will then place a tall red or green hat on each individual in line. You and your coworkers will be able to see the color of all the hats that are in front of you, but you will not be able to see the color of your own hat or the hat color of any individual behind you in line.

Once all the hats are fitted on their respective heads, each person will be given a clicker with two buttons, one red and the other green. The CEO will then approach the individual who is at the back of the line (the one who is able to see all the hat colors of all the other colleagues) and ask that person, "What color is your hat?!" That last person in line will then press the button that represents the answer, and a recorded voice over a sound system will speak the response "RED" or "GREEN." The CEO will then react with a loud, "You're right, get back to work!" or "You're wrong, YOU'RE FIRED!" The CEO will then move to the next person (the one standing immediately in front of the employee who was just yelled at) and will repeat the process until all employees have answered the question and discovered their fate, in turn. Everyone can hear the sound system reveal an individual's answer as well as the CEO's response.

Knowing what awaits you and your coworkers tomorrow, you all gather after work to devise a plan so that the fewest number of individuals will be fired. Create a scheme that can be used by you and your colleagues to save the jobs of as many people as possible. How many can you *guarantee* will not lose their jobs? *Note:* No "cheating" is allowed; that is, no additional information can be conveyed through additional words, the timing of the responses, gestures, et cetera.

6

The Prompts

Turning Puzzles Over in Your Mind

These prompts are mindsets that flow from practicing effective thinking. Many of these insights are from students enrolled in my *Effective Thinking through Creative Puzzle-Solving* class or from participants who attended one of my workshops on leadership, creativity, or teaching. The prompts themselves ideally could also be used as metaphors for thinking through all of life's puzzles. So the next time you are turning over a challenge in your mind, you are invited to turn over this book, read a few of these prompts, and apply some effective thinking to the issue at hand to engineer an original insight.

You will notice some recurring elements within these prompts. In some sense, that is the point: The number of variations on a theme that a composer could create is limited only by that artist's imagination. Similarly, the possible variations on just a few elements of effective thinking are equally endless. Thus, the reiteration should be read as reinforcing rather than repetitive.

Finally, you will see that none of these prompts require any special insights or understanding into the puzzle at hand or its surrounding context. All that is required for success is intentionally following a mindset that embraces the practices of effective thinking together with the patience to allow that mindset to carry you somewhere. That is, just as Seamus from Chapter 3, with just those practices and nothing more, you can provoke new thoughts, realize deeper understanding, and make greater meaning.

I.1. Who's Who?

To understand simple things deeply, don't ignore details or overlook facts. Whenever you are able, consider all possibilities, and follow the flow of ideas by letting each possibility play out to its logical conclusion and see where it leads. If it leads to a dead end, then discover why that potential possibility under consideration is, in fact, not possible. Once you have done so, you have failed effectively—you have made progress. Now consider another possibility and repeat.

In this case, start by unpacking the meaning of the phrase "at least one of these students is lying," but more importantly, practice these effective mindsets beyond this puzzle.

I.2. When Six Equals Eight.

To understand simple things deeply, create easier questions, answer them, and then put forth the effort to transform those answers into a new insight—intentionally open the floodgates and be carried away by the flow of ideas. Take a failed attempt (intentional or otherwise) and study variations on the theme— wonder: "That didn't work, how can I modify that previous attempt?"

In this case, you can make the simple observation that three line segments are required to make one triangle, and if the six line segments do not cross themselves, you can draw only two triangles.

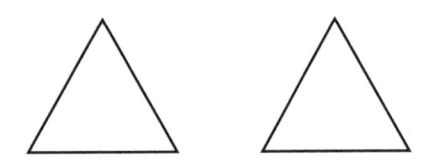

Thus, you are led to the simple, but essential, conclusion that the line segments must intersect. This basic insight leads to a new question: *How can two equally sized, equilateral triangles intersect each other?* There are at least two ways, as illustrated here.

Now consider variations, such as changing the relative position of the triangles or extending the line segments to six longer segments still of equal length. Alternatively, you could intentionally fail by returning to the original question but now removing the constraint of having the line lengths be equal. Now extending the previous two-triangle experiments with different-sized triangles, you would find several new configurations, for example:

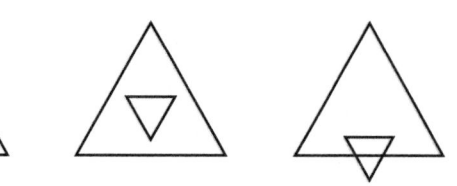

You could now ask: *What happens if I extend all line segments to lines of equal length?* See how many different successful configurations you can discover by seeing the flow of ideas of these failed attempts arising from answering easier questions; but more importantly, practice this triangulated effective mindset in another context.

I.3. Cutting Boards.

To deeply understand the big picture, it often is important to make meaning at the micro-level to uncover structure or a pattern. The practice of adding the adjective is always a sure way to provoke such thought. *The challenge is to keep adding adjectives and other descriptors until some subtle, simple, or otherwise hidden structure is revealed. By doing so, you will be able to see what had been squarely in front of you but that otherwise would have remained invisible. Another way to go deep at the micro-level is to simplify a scenario to the most trivial case possible. Practicing effective thinking empowers you to make the invisible visible as well as meaningful.*

In this case, consider a domino placed, as described in the puzzle, on any of the three chessboards. Now add the adjective to express how that domino covers the chessboard and what it covers. Through that exercise, expose some subtle difference between the first two chessboards as compared to the last one, but more importantly, practice these mindful moves elsewhere. Alternatively, consider shrinking the dimensions of each of the three chessboards to as small as possible while still capturing the distinctive features of each, and then considering the domino covering question with those simpler scenarios.

II.1. Three Switches, Two Rooms, and One Bulb.

Often an optimal solution can be uncovered only by first finding any—even a ridiculous—solution and then refining your thinking. So don't stare at a blank screen. Instead, pro-

duce something *and then use it as a starting point from which to respond—refining your thinking refines your final product.*

In this case, it is easy to see that at most three hall trips would suffice: Turn one switch on, walk down the hallway to see if the light is on, and if it is off, walk down the hallway a second time to return to the switches, turn that first switch off and the second switch on, and walk down the hallway a third time. No matter if the desk lamp is on or off, you know which switch controls it. Now you could create questions to provoke a further refined thought: *Can I come to a definitive conclusion requiring fewer trips? Isn't the only way to determine if a switch controls a light to look and see if the light is on when the switch is on?* This last question can lead to new insights as well as a bizarre, alternative question: *Could one determine the correct switch if one were blindfolded?* Now you have moved from thinking about minimizing the number of trips down the winding hallway to a related but different challenge. Even if there were a way to determine the correct switch while blindfolded, it would require more than a few trips; so you are assured of failure. But, perhaps by intentionally failing, you will have an illuminating "Ah-ha!" moment. Use these mindsets not only to enlighten the way you look at these puzzles, but also to enlighten the pathways to the challenges to come.

II.2. Going from 5 to 4 in Two Moves.

When in doubt (and otherwise) challenge yourself to add the adjective and keep doing so until something new is exposed. Only when you explicitly describe something can you discover structure and uncover a pattern.

In this case, keep describing the given collection of matches until you gain an insight into the relative positions of the four squares that need to be created. Then see what's there and see what's missing, and finally use these observations to look for a pattern . . . not just within these squares, but within your life.

II.3. A Slow Burn.

Remember the adage: When the going gets tough, the smart do something else by creating easier challenges, resolving them, and repeating that process until a new insight is realized.

In this puzzle, the challenge at hand is far too difficult, so instead of burning out, ignite some intermediate success. Certainly you could, by burning, measure out 60 minutes and even 120 minutes. Now you are led to ask: *Are there any other times I could measure precisely in this way?* To get different time lengths, you must apply different processes. So what are some variations on the theme used to generate 60 or 120 minutes? Recall that cutting the strings will be of no help: The burn rate throughout each string is not uniform. Once you discover all the time periods you can measure without too much difficulty, see if you can recycle your creative ideas to generate a fresh insight and, consequently, measure 45 minutes. Creating easier questions, answering them, and repeating until a realization is revealed is a mindful practice for resolving issues large and small, even those with no strings attached; so apply it freely and often to ignite the flame of deeper understanding in your life.

III.1. A Top 10 List.

Often the best way to make deeper meaning is to look at an issue from different perspectives; and, if you do not know where to start, begin by considering the opposite point of view.

In this case, hidden structure might be revealed by rephrasing each sentence as a logically equivalent assertion about how many statements are true, that is, "Exactly *xx* statements on this list are true." What new insight or thought is provoked by looking at this puzzle from this new perspective? Practice this lesson of developing a deeper understanding by looking at an issue from the opposite side throughout your life.

III.2. Five Elements, but Only Four Hats.

Consider alternative scenarios and ask if they lead to alternative outcomes. Related, but of independent value, see the world from different vantage points. Different perspectives often lead to greater empathy and deeper understanding. But to have an epiphany, you need to follow the flow of thinking from that perspective until it leads to an insight; otherwise you are not truly seeing the world clearly from that angle. You need to think through that point of view—understanding that mindset and the meaning of any observed actions or lack thereof.

In this case, all four students know that two hats are black and the other two are gold. But do all students have the same information or do some students know more about this puzzling situation than others? Consider alternatives: *What would have happened if the hats were placed in a different order? Why did it take a minute or so for one of those smart students to call out the correct hat color?* Deep empathy into different vantage points can change how you see the world—practice this effective mindset beyond this hatful puzzle to discover inner peace.

III.3. Roommates from Ruter.

This puzzle celebrates the basic practice of understanding simple things deeply. When faced with a truly difficult chal-

lenge, such as this one, remember: *Don't do it!* Instead, create a lesser challenge—in fact, create the simplest related challenge you are able. Resolve that easy puzzle and understand its solution in unusual depth. Create a slightly less simple version of the original challenge, resolve that slightly more challenging puzzle, and understand the connection between your two solutions. Repeat to discover some pattern that will carry you back to see the original, daunting challenge with greater nuance. That clarity can help uncover some otherwise hidden structure that will lead to the solution to the original challenge.

With the puzzle at hand, you might first want to understand the situation a bit more deeply by asking: *Given the rules of the Ruter riddle and knowing that Ralph heard nine different answers, what were the numbers Ralph heard as responses? Could someone have shaken nine hands?* (Recall that no one shakes his own hand or his roommate's hand.) With this answer in hand, you could wonder what part of this party conundrum makes it so complicated. Once you generate a helpful response, you could then ask: *How could you keep the rules and essence of the puzzle intact but create a simpler yet similar scenario?* Ask this question throughout life to shake loose new insights.

IV.1. A Dismal Downpour.

This silly riddle celebrates the lesson that even when you are drenched in details, no fact should be overlooked—some tiny feature might shed the necessary light to illuminate the entire issue. Also, thinking through a situation or story from the beginning—including running through its entirety in your mind—and visualizing the circumstances unfold until the conclusion allows you to see the entire picture.

In this case, the action starts at the stroke of midnight, but effective thinking can be applied at any hour of the day.

IV.2. Making 4 from 12.

We often hear "think outside the box," but that suggestion does not tell us how to move our minds beyond the boundaries of the box. The practices of effective thinking offer practical actions that you can take to provoke thought (even beyond the box). Creating easier or alternative questions and looking at extreme cases are two straightforward pathways that lead to outside-the-box thinking.

In this case, the rightmost figure in the puzzle (the polygonal plus sign having area equal to 5) suggests that you need to think outside the square—that is, you might not be able to produce a polygon with area 4 having only right angles. What does this insight imply? An alternative warm-up question is: *What's the largest number of matches I could use to make a polygon having an area of 4 in which all the corners are right angles? Could I then expand that figure to a figure with the same area but one that includes any remaining matches?* Yet another approach to this puzzle is to ask: *What's the smallest area of a polygon that can be constructed using the 12 matches?* Because a polygon must have an inside, there must always be some area, so zero is not possible. But can you create a polygon having any small positive number as its area using the 12 matches? If so, can you apply the thinking behind your answers to solve the original puzzle? In this direction, modifying the leftmost figure (the 3×3 square) might provide some answers. Often answering a challenge head-on is the wrong approach—instead come at it askew through a chain of different but related questions that lead to new ways to see the original challenge from a new angle—one that ideally makes its solution easy. Now apply this mindset to an issue beyond this puzzle.

IV.3. Map Folding.

One way to understand simple things deeply is to consider an issue at the micro-level, ignore the distraction of the entire challenge, and attempt to make incremental progress. *Moving to the macro-level often requires an entirely different perspective. Combining the insights from both micro- and macro-thinking can generate an epiphany or, ideally, a solution. Finally, actively playing with ideas hands-on, whenever possible, is a powerful way to stumble upon discovery. Tenacity is required as the stumbling—sometimes even just trial and error—results in intermediate, but ideally effective, failure. Whenever possible, turn that intermediate failure into an insight before moving on to the next attempt. Incremental thinking offers the promise of a breakthrough idea. Remember the great wisdom of Winston Churchill, who might have said:* "Success consists of going from failure to failure without loss of enthusiasm."

In this case, for micro-thinking, consider one pair of consecutive numbers at a time, and for each pair, find a way to fold the map so that their corresponding squares touch. If getting the two squares to match up for a particular pair of consecutive numbers seems impossible, identify the constraints that appear to prevent that pair from touching and focus on overcoming those constraints. If overcoming that identified constraint does not seem possible at that moment, identify why, and consider the next pair of consecutive numbers and repeat. Once you have two squares aligned correctly, try to expand your thinking to adjacent numbers. For macro-thinking, consider all the ways that the map could be folded along its lines. There are some truly counterintuitive folding schemes possible. So exploring different ways of folding, with the constraint of the numbers removed, is a liberating way to devise intricate folding schemes that might be useful when you return to the micro-level and reintroduce the number-adjacencies constraint. But more importantly, practice this lifting of constraints together with balancing micro- and macro-thinking to the ever-unfolding issues in your life.

V.1. Whoops.

Often in the puzzles of our lives, we get so bogged down with the details that we miss the big picture. Details are certainly important, but they should never eclipse the entire situation. Effective thinking requires looking at challenges with the mind's eye through a telescope as well as a microscope.

In this case, keep rereading the entire puzzle as carefully as possible, but more importantly, identify the details hidden within the other puzzles of your life.

V.2. A Not-so Weighty Challenge.

Incremental progress is always progress, and often a divide-and-conquer approach, when repeated, will carry you to a smart resolution of an issue. Alternatively, considering an easier question always offers an insightful lift.

With this puzzle, instead of identifying the hidden gem within the group of nine, could you first identify a smaller group in which the gem is present? There are many ways of dividing the totality of stones into smaller groups, and each possibility should be considered until one way leads to certain identification of the gem with just two weighings on the balance scales. In addition, you might consider an easier question: *Could you answer the question if there were only two stones, one of which contained the gem? What about three stones? Four? Five?* These toy questions might offer some useful footholds as you make the ascent to resolve the original nine-stone puzzle, but creating toy questions always helps lighten life's weighty issues.

V.3. A Star Is Born.

Tenacity, together with considering a different but connected question, is a helpful way of realizing deeper understanding.

In the case of this very challenging puzzle, I must confess that when I was first faced with this conundrum, I could not figure it out—I used it as my "go-to puzzle" while on airplane trips and invested many hours of flight time drawing star after star, trying to unravel this riddle. I'm sure that the people sitting next to me questioned my mental state. When I finally figured it out, I felt great, but soon was disappointed that I hadn't asked a different question from the very start. I now question my mental state all the time by creating different questions whenever possible.

VI.1. Puzzling Politicians.

To resolve any issue, you must first understand it deeply and as fully as possible. Often adding the adjective allows you to make even greater meaning of that with which you are faced.

In this case, we are given four facts. Add the adjective to each—that is, describe each fact in as much detail as possible and express each fact in as many different ways as possible. Those additional descriptions will add the color and texture to help you see the panorama of the entire situation in greater detail and to realize the resolution—not just within the puzzles of politics but throughout life.

VI.2. Turning on a Dime.

Divide and conquer is always a useful tool to engineer an insight.

With this puzzle, notice that the "glass" is really a "T" with two additional vertical arms. A first step might be to wonder: How can you turn the glass upside down (and in how many different ways)? To turn the glass upside down requires that the "T" appear upside down. So as a start, focus your efforts on the intermediate challenge of creating as many different ways of constructing an upside-down "T"; then, for each construction, consider the vertical arms. Now try dividing and conquering in order to find creative solutions that fit each issue in life to a T.

VI.3. Penny for Your Thoughts.

Whenever possible, apply all five elements of effective thinking to a daunting challenge. Often it is easy to fail, but far more difficult to fail effectively. After investing the time to deeply understand the challenge at hand, you might want to consider easier cases. Remember, with these easy "warm-up" scenarios, success is not in finding an ad hoc *method that works only for that special case, but rather in discovering an otherwise hidden pattern and general structure—that is, these warm-up questions should provoke an idea that flows to cases beyond the easy ones.*

In this case, after taking the time to truly understand the challenge itself, you may want to consider easy cases: *What if I had only one penny on the table? Two cents? Three cents?* Challenge yourself to transform an *ad hoc* method that works for that special case to a more general process that can be extended. You might also find it helpful to consider some special cases, for example:

What if one of my two collections contains all the heads-up pennies? What if one of my two collections contains all but one of the heads-up pennies and one heads-down penny? Considering extreme scenarios can often lead to a new insight.

VII.1. A Doggone Puzzle with Cats.

A pet strategy for intentionally failing is considering extreme cases, asking probing questions, and patiently seeing where those efforts lead and what is learned.

In this case, assume that all ten animals are the same type of pet and react to that certainly wrong answer: *How many biscuits have I used? How many are unused? What could I do with each of those extra biscuits?* What new insight are you led to about this puzzle?

VII.2. A Cross Farmer Needs to Cross a River.

To understand deeply you need to force yourself to see the otherwise invisible by considering the counterintuitive.

In this case, you could focus your thinking on not only on the order in which the farmer brings over the three, but also on what Francis does before heading back to the starting side of the river.

VII.3. Making 50/50 Greater than 50/50.

Consider an easier question to build deeper understanding; and if no new insight is generated or structure discovered, create another question and repeat.

For this puzzle, consider the easier question in which you only have two black marbles and two gold marbles. In this scenario, you could write out all possible ways of placing the four marbles in the two bowls (there are only five different ways), and see which one optimizes the likelihood of selecting a black marble while blindfolded. Always create easier questions or simpler scenarios to gain greater clarity into the more complex.

VIII.1. The Deficient Digit.

By asking basic questions and probing their fundamental answers, you can often uncover subtlety, nuance, and differences—that is, understand more deeply. Also, looking at simple cases first can also offer an epiphany.

An easy way to launch this type of thinking is to consider a simpler situation. (*Why look at a thousand numbers first?*) Alternatively, one might naturally believe that all ten digits are created equal—that is, that they all have the same properties. However, this puzzle implies that is not the case. That surprise could lead you to ask:

Which digit is different from the others in terms of using it to generate numbers from 1 to 1000? Looking for differences and distinctive features at a very basic level allows you to understand simple things deeply as well as discover important previously missed insights.

VIII.2. Going in Circles.

One way to understand more deeply and engineer an epiphany is to consider a specific example of a generic circumstance. Analyze and think through that case in order to see the more general situation in a different way. Also, creating new and easier questions always jump-starts an otherwise immobile mind.

Here, suppose that last year's graduating class was made up of 500 students not owning cars and 500 students with cars. Further suppose that the non–car owners had an average GPA of 3.0 and the car owners had an average GPA of 2.0. Can you create a new graduating class of 1000 students so that the average GPA of each of these two groups is higher than the corresponding average of the previous class, but the overall average GPA of all 1000 of those in this new class is relatively lower than that of the previous class? What new insight about averages and statistics is revealed through your attempt?

VIII.3. Slicing a Square in Half in Three.

Applying the flow of ideas often, if not always, leads to some fresh ways of looking at the situation. Taking a previous solution, insight, or epiphany and trying to reimagine it in a new context can be the inspirational spark to generate a new idea. Creating easier questions might also allow your creativity to flow even further.

In this case, there are a number of earlier puzzles involving matchsticks. Could an insight from one of those be extended to the puzzle at hand? As for a different, easier question, you could ask the warm-up: *How can I arrange three matches (touching end-to-end) so that they merely fit inside the 2 × 2 square made from eight matches?* Thinking through an answer to an incremental, warm-up question might catapult you to a solution to the original challenge at hand.

AN EXCEPTIONALLY CHALLENGING BONUS
Practice Effective Thinking through . . .

The **C**razy and **E**xceptionally **O**bnoxious **CEO**.

One way to understand more deeply is, whenever possible, to articulate clearly all that you know and all that remains unknown.

Understanding deeply in this context begins with the realization that when the CEO approaches you for your answer (no matter your position in the line), you know *every* hat color of each of your colleagues—by listening to the CEO's reaction to those behind you on line and by looking at the hat colors of those ahead of you in line. Thus, at the moment you are asked for your response, you know everything *except* your own hat color. This insight is significant and might be overlooked on first inspection. Now imagine you are the person in the back of the line, who is asked to answer first. Is there anything you can convey through your answer to provide the coworkers ahead of you in line with an insight that only you alone have from your vantage point? As always, considering special and easy cases first can lead to insights as you extend your thinking to the more general or complicated challenge at hand. In this case, what if you had two other coworkers? Three? Four?

7

The Insights

Reflections on Thinking through Puzzles

In this chapter, selected epiphanies that could be gleaned from practices of effective thinking suggested by the previous chapter's prompts are outlined. Those insights change how we see the puzzles. That different and deeper view of the puzzle is always right in front of us—but it requires effective thinking to lift the initial fog and see the puzzle in a clearer way. By provoking our own thought, we can make the invisible come into sharp focus in our mind's eye. I hope you find these commentaries worthy of further and future reflection.

I.1. Who's Who?

The fact that one student is a math major and the other is a philosophy major together with the fact that at least one of them is lying leads to only one possibility: They both must be lying.

Unpacking important details and understanding simple things deeply will allow you to open your mind to all possibilities on future issues.

I.2. When Six Equals Eight.

In this case, extending the six line segments to six longer line segments of equal lengths will transform each of the earlier five failed attempts into a successful configuration. For example, if we start with

and extend all six lines, we would generate

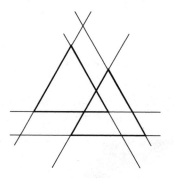

which contains eight equilateral triangles. Furthermore, if we just change the relative positions of the second failed attempt by moving the top triangle downward, we would discover another solution, namely, a Star of David. The interesting point here is that every naive, initially failed attempt leads to success.

Notice how you can, with intentionality, create a multiplicity of solutions to life's puzzles by practicing the mindsets of considering simpler issues, creating answers that are known not to work and extending those failed attempts to generate new discoveries or insights—not just with a crossed line but anytime you find yourself at an intellectual crossroads.

I.3. Cutting Boards.

A domino correctly placed on any of these chessboards might be described as: Covering two squares; if pushed further, as: Covering two adjacent squares; even further: Covering two adjacent squares having different colors. Hence an insight is revealed: A correctly placed domino must cover one black and one white square. Thus, whenever you can cover a chessboard (cut, truncated, or untouched) with dominos following the rules of this puzzle, you discover a fact about the number of black squares as compared to the number of white squares.

You could also come to this essential insight by considering 2 × 2 chessboards, in which case the observation is easy to see. Either way, this insight resolves the puzzle. Notice how through these intentional mindful practices, you see the three chessboards with deeper clarity—you actually look at them and see new detail that was probably missed on first inspection.

Apply these same practices to see something in your life in a more focused way to uncover some otherwise hidden nuance.

II.1. Three Switches, Two Rooms, and One Bulb.

If one is patient and uses another of the five senses beyond sight, one can resolve this real-world puzzle about a real-world desk lamp in a surprisingly few number of trips down the hallway. In our world today, we often want to proceed as quickly as possible. However, time is a variable that can be a useful tool in being more effective—sometimes there is value in waiting before taking an action. But realizing such opportunities requires deep understanding by simply proposing a plan and then reacting to that less-than optimal proposal by asking many questions—some of which might turn the original issue on its head to visualize the entire situation in a new light.

Now set your mind to another issue and slow down to apply the same effective thinking to have a light bulb go on in your head.

II.2. Going from 5 to 4 in Two Moves.

There are sixteen matches that are to be used to create exactly four squares. The new insight revealed by these two adjectives is that we can have no match be a common side to any two squares. Thus, we must have no adjacent squares in our modified figure. Once you discover the solution, you will see it connect with a puzzle from Challenge I and have a Paideia moment.

Whenever you move from a qualitative perspective to a quantitative one, you are, in fact, adding the adjective and creating an opportunity to lift the cloud of superficial and fuzzy understanding and see the structure and pattern that was previously obscured. Now set your mind to another issue and clear away the clouds that block deep understanding.

II.3. A Slow Burn.

Could you measure exactly 30 minutes by burning the strings? Looking at an issue from both ends of the spectrum can offer a fresh perspective and generate new insights.

Once you have a new insight, consider the flow of ideas—that is, wonder how that insight can be reapplied to realize something even bigger ... not just how to burn up 45 minutes, but how to fire up your mind to ignite new ideas throughout life.

III.1. A Top 10 List.

Could two of those sentences be true? Given the answer to this basic question and looking at the statements from a different perspective reveals the answer to the puzzle itself.

Do you see the glass as half empty or half full? There is value in viewing the glass through each of those two opposite, albeit equivalent, perspectives. In doing so, you not only resolve this riddle but also see the world around you with greater awareness.

III.2. Five Elements, but Only Four Hats.

No one speaks for a minute. Certainly we would not expect Alicia or Bryce to say anything since they see no hats. On the other hand, Devin sees the greatest number of hats. Carol, who knows all of these facts, sees only a gold hat on Bryce. Carol then wonders what Devin might see from his vantage point and goes through all the possibilities—some of which would allow Devin quickly to figure out his own hat color. But he said nothing.

The view from another perspective and the facts at hand (or on head) speak volumes as to why there was silence for a span of time. Now put on your effective thinking cap and allow time to inspire your own thinking.

III.3. Roommates from Ruter.

The most surprising aspect of this shaky puzzle is that it can be answered at all. The second-most surprising feature of this puzzle is that by understanding simple things deeply, the answer appears in the palm of your hand. The challenge here is that there are so many people shaking so many hands. The simplest version of this scenario is the case in which they invited no other pairs of roommates. In that case, it would be just Ralph and Rich and, given the rules, Rich would shake zero hands. The next simplest case would be if they invited one other pair of roommates, say Adam and Bill. When Ralph asks the three for the number of hands they each shook, he hears three different numbers, and those numbers must be 0, 1, and 2.

The puzzle remains challenging, so it might be helpful to create a new question: Given the simple scenario at hand, can we determine how many hands are shaken by Adam and by Bill? Notice that this different question would automatically give us the number of hands Rich shook, but it is framed in a different way. The numbers of shakes in this special case are small enough that you could consider all possibilities: if Adam shook no hands and Bill shook one; if Adam shook no hands and Bill shook two; if Adam shook one hand and Bill shook two; et cetera. The answer will help uncover not only a new insight but also a pattern as the number of roommate-pairs at the party grows and thus a firm grip on a way to answer the original puzzle.

Deeper understanding follows from the structure discovered in the seemingly-simple.

IV.1. A Dismal Downpour.

If an additional epiphany is required, you might find it useful to convert 72 hours into days.

Even a silly puzzle can amplify your thinking on the far more serious issues in your sunny life.

IV.2. Making 4 from 12.

The simplest polygon having an area of 4 using only right-angles is a 1 × 4 rectangle.

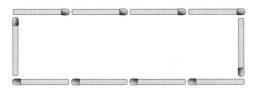

This rectangle is made up of 10 matches, so the question is how to modify the figure to include the remaining two matches.

Alternatively, leaving this line of thinking for the moment, you could consider the smallest area of a polygon you could construct using 12 matches. If you consider the 3 × 3 square (as in the leftmost figure of the puzzle) having hinged corners, you discover that you could deform the square by sliding the upper horizontal edge to the right and keeping the bottom edge fixed to form a drunken square, also known as a rhombus.

If you let the top edge of the rhombus collapse onto the bottom edge, you would generate a degenerate polygon in which two rows of six matches are on top of each other. That degenerate polygon has area equal to 0 and is really not a polygon. However, if you lift that top side just a bit, you would have a diamond-like sliver of area. The area of the original square is 9 (3 × 3) and the area of the drunken square could be made as small as you wish. What could you now conclude regarding the original challenge?

The hinge idea could also be used in the first attempt with the 1 × 4 rectangle: Replace the right and left edges by two matches forming a "V"

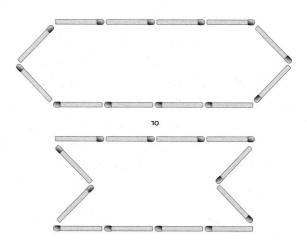

or

and then apply the same reasoning as before to show there must exist some polygon made of 12 matches and having an area equal to 4.

There are many different solutions that you can find once you loosen constraints and consider what happens when you lift the assumption of rigidity and consider a more malleable possibility. Agile thinking can lead to shaping any issue in a totally new dimension—apply these mindsets to unhinge another of life's puzzles.

IV.3. Map Folding.

For the first map, consider all ways of bringing the "7" and "8" squares together. For each such folding, see if square 6 can "join the fold." Work out from there. For the second map, an unusual folding scheme is required at the end. Create a folding scheme that is inspired by the act of turning a tube inside out.

Now set your mind to an origamically unrelated challenge and apply the same effective thinking used in this map puzzle as an intellectual GPS to drive you to an uncharted solution.

V.1. Whoops.

Every number in this riddle is important. Every single one.

Attention to detail opens pathways for effective thinking.

V.2. A Not-So Weighty Challenge.

My students initially try to weigh four stones against four stones, but quickly realize that this procedure will not always guarantee finding the pirate's booty with just two weighings on the balance scales. That failed attempt, however, carries them to an insight that allows them to discover the treasure.

Weigh other issues with similar mindsets to uncover new riches like a bright Buccaneer.

V.3. A Star Is Born.

Instead of addressing the question of creating as many triangles as possible, one might ask a different one: How could two lines be placed over the star so that they cross the star as much as possible? That is, instead of maximizing the number of triangles, try to maximize the number of line crossings.

Another point of interest: I named the puzzle this way because I thought it was a cute title. A few of my smart Southwestern students took it as a hidden hint and it led them to the solution. They observed that in their solutions, they introduced a new smaller star ... so within the solution, a star is born. Brilliant!

Considering a different question can allow you to reach for the stars.

VI.1. Puzzling Politicians.

"At least one of the politicians was honest" means that one or more is honest. "Given any two of these politicians," means that you may consider any two from that group—which includes the possibility of selecting one and then pairing that one with the remaining politicians, one pair at a time.

Add the adjective to come up with an honest-to-goodness new insight in any context.

VI.2. Turning on a Dime.

Even though sliding a match is certainly a "move," it is something often overlooked when one considers "moving a match."

Intentionally becoming aware of simple possibilities allows opportunities otherwise unforeseen far beyond upside-down puzzles—try it as you contemplate your next right-side-up challenge.

VI.3. Penny for Your Thoughts.

The only information you are given is the number of pennies lying heads-up. That number should somehow inform how you divide the pennies into two collections. You might then consider variations on the theme—in this case, an alternative scenario—and compare the number of heads-up pennies in one collection to the number of heads-down pennies in the other.

Whenever possible, see how many of the elements of effective thinking you can apply to expose hidden nuance—try applying all five to engineer that inevitable epiphany and land heads-up on a clever solution.

VII.1. A Doggone Puzzle with Cats.

The incorrect assumption that all ten pets are cats leads to the realization that, at least in the context of this puzzle, a dog is nothing more than a cat with an extra biscuit. This basic but important insight not only allows you to see the original puzzle in an entirely new light, but also immediately reveals its answer.

Take a moment or two to appreciate how you have changed: You see this puzzle differently now than when you initially read it. As a result of that change in understanding, you can now quickly answer generalizations or variations of the puzzle. For example, what if you have 20 pets and 111 biscuits?

Congratulations—change through effective thought offers the exciting promise of growth. Whether you're a cat or dog person, see a pet issue of yours in a new light by feeding your mind first to consider extreme perspectives.

VII.2. A Cross Farmer Needs to Cross a River.

By bringing some previously transported cargo back across the river, Francis can deliver all three safely to the other side.

More importantly, considering the counterintuitive—that is, all possibilities—allows you to see what otherwise would have been invisible in your mind's eye.

VII.3. Making 50/50 Greater than 50/50.

We need to average the likelihood of selecting a black marble from each bowl. So if one bowl contains two gold marbles and the other contains two black marbles, then the likelihood of winning if you select a marble from the first bowl is 0%, and the likelihood of winning if you select a marble from the second bowl is 100%. Averaging these two likelihoods produces a 50% chance of winning. Now can the four marbles be placed in the bowls so that the likelihood of winning is greater than 50%? How can you extend your thinking to the original 100-marble scenario?

When effectively organizing the marbles in your head, place yourself in a position to engineer the flow of ideas.

VIII.1. The Deficient Digit.

Repeating a digit (such as 2, 22, 222 or 7, 77, 777) generates different numbers except in one case. That insight should lead to the deficient digit. As for the remaining nine digits, they all appear with the same frequency as each other in the numbers from 1 to 999. However, when you also consider the number 1000, you have an extra digit of 1 appearing. You might have noticed this phenomenon if you first considered this challenge with the numbers 1 to 10.

Ask basic questions and find nuance in what otherwise first appeared too simple—you will see a complex issue in a deeper way.

VIII.2. Going in Circles.

The insight here is that knowing the relative sizes of sub-populations is essential if you wish to make meaning of comparative averages and statistics. In this case, suppose that the new graduating class is comprised of 100 non–car owners with an average GPA of 3.1 and 900 car owners with an average GPA of 2.1.

Effective thinking—in this case, considering a specific situation—empowers you to process information more deeply and make true meaning of information and data.

VIII.3. Slicing a Square in Half in Three.

Position three matches end-to-end and so that they meet at right angles—there are two such configurations. Consider one of these configurations should lead to a solution to the puzzle. To answer the bonus, suppose there were hinges on those right angles and slightly straighten out the previous configuration.

Whenever possible, recycle ideas, insights, mindsets, and, of course, elements of effective thinking to future issues, scenarios, and challenges you will face.

AN EXCEPTIONALLY CHALLENGING BONUS
Practice Effective Thinking through . . .

The Crazy and Exceptionally Obnoxious CEO.

If you are the person in the back of the line, you must give only one of two responses. So looking at the hat colors of all your colleagues standing in front of you, how can you describe what you see in one of two possible words? A great guess is to say the hat color that appears with the greatest frequency. This attempt is problematic, for example, if the number of red hats equals the number of green hats. With that failed attempt as an intermediate inspiration, you discover you need to be more specific. Return to the question and consider it more deeply: Looking at the hat colors of all your colleagues standing in front of you, how can you describe what you see? *Now, as always when trying to understand more deeply, add the adjective.*

8

The Future Has Arrived . . .

Well, my dear reader, if you have made it to here, and read all that came before, including the upside-down and mirrored words of the previous two chapters, then you have traveled far . . . and further still, if you took the time to actively engage and practice effective thinking for yourself. In that case, you have experienced some of the intellectual journey mapped out in my course *Effective Thinking through Creative Puzzle-Solving.*

At the heart of high-impact, formal education is not what to think about any particular subject or individual issue, but rather how to think effectively through everything—and practice that thinking everywhere.

A tradition I proudly introduced at Southwestern University is one in which before graduation I invite every student to the president's house for dinner at the president's table—an impressive table that seats eighteen diners: twelve students and six faculty, staff, trustees, alumni, or friends of the university.

As dessert is served at this formal meal, I call the many lively conversations to a close and introduce a previously unannounced topic for the entire group to engage in one conversation—and everyone contributes, which is also a metaphor for the culture we should always foster. You can find those topics (along with group pictures around the table) on Twitter by searching for @ebb663 and the word "dinner." If you are so moved to host your own such events, feel free to use this book or previous president's table dinner topics as prompts for your own engaging conversations. Connecting with others through such uplifting conversations is a powerful way to provoke further thought and continue to practice effective thinking.

Congratulations again, and may you not only continue to wisely learn, grow, think, and create, but also, in the true tradition of Southwestern's *Paideia* philosophy, continue to connect ideas. In doing so, you will offer yourself a truly meaningful and enriching education today with the promise of an even brighter tomorrow.

APPENDIX

An Expanded Overview of the Course

In the fall of 2015, I created the course *Effective Thinking through Creative Puzzle-Solving*, although transcripts list it as *Effective Thinking through Creative Problem-Solving*, as employers value clever problem solvers and might not see their challenges as the puzzles they truly are. As noted previously, in reality, we all face puzzles every day in our lives—personal, professional, small, and existential ones. Some of them can be cast in the negative light of *problems*, but life's puzzles extend far beyond life's problems. I outline the course here so that readers can generate similar experiences for themselves and for others.

Although this class is the most profound course I have ever taught, as mentioned earlier, I call it the "*Seinfeld* of the curriculum," because the course is about nothing yet attempts to teach everything. It contains no short-term content; instead there is only the long-term goal—offering the answer to what I call The Teacher's 20-Year Question: *Twenty years from today, what will my students still have with them from their experience in my class?* I hope that my students will joyfully practice mindsets that will lift their creativity, enhance their abilities to make connections, and strengthen their aptitudes to think effectively throughout their lives.

Instead of content focused on a topic, such as calculus, the content of this course focuses on the learners' minds. We discuss the physiological development of the brain, the negative impacts of multitasking, social media, and personal electronic devices, as well as the positive impacts of mindfulness, gratitude, and other ways of allowing one's brain to recharge as well as to focus more deeply. But the core of the course is centered on specific practices of effective thinking that empower individuals to think in new ways, become more creative, and discover connections that otherwise would have remained invisible.

These mindsets are practiced through a sequence of puzzles—three each week: one relatively straightforward, one a bit more difficult, and one that is intended to be truly challenging. All, however, are designed to provoke thought; the ultimate goal is not to solve the riddle at hand, but rather to apply multiple practices of effective thinking to see that puzzle in as many different ways as possible even after a solution is discovered.

Solving the puzzle is like receiving the diploma—it's not the thing. The mindful journey that led to an imaginative insight or solution is the thing. That journey will promote the intellectual agility to see features of our world one way and then apply practices of effective thinking to see those same features in a brighter light. However, moving from that natural temptation to search for a quick solution to an enlightened desire to embark patiently on a thoughtful journey and discover where it leads is extremely challenging without much practice. These puzzles offer the practice needed to make that enlightened perspective less challenging and more natural.

Beyond thinking through the puzzles of this class, students also are expected to apply the mindsets from our course to the rest of their lives: Students submit copies of homework assignments from their other classes and highlight the elements of effective thinking applied to enhance their final products. They also apply those mindful practices to solving puzzles beyond their courses: They improve their relationships with family, friends, coworkers, or roommates; they strengthen their strategic abilities on sports fields or in the workplace; and they amplify their efforts in all of life's extracurricular activities.

As the course is about thinking and the mind, students are encouraged to practice a form of mindfulness of their own choosing—it could be through a quiet walk, sitting still, expressing gratitude, meditation, or any other activity in which the mind is allowed to rest in an awakened state in which the focus is on the *now*. So often our minds are in the past, troubled by what had just happened ("I really messed up; I'm in big trouble"), or in the future, distressed about what is to come ("I'm so unprepared; I'm in big trouble"). In our technological world, with electronic distractions ever pinging in our pockets and bags, we tend not to live in the moment—instead we are someplace else, or in many other places at once. Science shows us that the mind needs an opportunity to settle and recharge in order to operate as effectively, wisely, creatively, and joyfully as possible. Thus, embracing moments of solitude not only offers the promise of this personal growth, but also opens us to greater empathy in how we engage with each other.

Each week, an interesting guest visits the class for an interactive 90-minute session. The visitors open by briefly sharing the puzzles in their lives (professional or personal, large or small) together with the practices they employ to think through those puzzles. The rest of the class time is turned over to the students as they practice the art of creating questions—as each student asks these highly successful people a probing question. Those hour-long Q&A sessions always are the most dynamic and stimulating parts of the guests' visits.

Through all these and other facets of the course, students practice new mindsets in a variety of contexts.

And practice is the key word; even though many of the elements and exercises offered to provoke thought initially and deceptively appear straightforward. The challenge is to embrace them as one's own and incorporate them into one's natural patterns of creativity and everyday thought. To emphasize the importance of practicing the mindsets introduced, students are asked to sign a contract listing all the terms they agree to follow. The contract and first-day course materials are appended here.

I hope that as you move through this book and journey through your life, you too will commit to being open to new templates of thinking and modes of analysis—they hold the promise to carry you to even greater heights.

You are invited to engage in this unique curriculum and enjoy the intellectual stimulation fostered in this class for yourself. You can explore different forms of mindfulness, engage with interesting people around you and discover how they solve the puzzles in their lives, and you can practice effective thinking through the puzzles and prompts herein. Embracing the practices of an effective thinker requires individual agency and ownership. You cannot passively sit back in an entitled position and demand, "Educate me." Instead you must intentionally generate your own opportunities to challenge and, over time, transform yourself. Your schools, teachers, professors, mentors, and even this book can play at most supporting roles in the story of your intellectual journey—you remain the central character and hero in that thoughtful adventure that continues to unfold . . . that adventure known as life.

Effective Thinking through
Creative Puzzle-Solving

University Studies 232   Southwestern University

    Instructor: Dr. E. B. Burger

Texts:

- *Making Up Your Mind* (MUYM) by Edward B. Burger (Princeton University Press, 2019).
- *The 5 Elements of Effective Thinking* by E. B. Burger and M. Starbird (Princeton University Press, 2012).
- *The Little Prince* by Antoine de Saint-Exupéry (Mariner Books, 2000; originally published in French in 1943).

Description:

This two-credit experimental course will sharpen your problem-solving skills and your ability to create novel approaches and see issues from a variety of perspectives. You will also develop ways of understanding at a deeper level by engaging with different ideas across the curriculum and through logic puzzles and mind-benders. An intentional component of this course is to connect these mindfulness practices to your other classes and to the rest of your life. In addition, weekly special guests with a wide variety of intellectual interests will visit, share their ways of thinking and their life stories, and engage in thought-provoking conversations.

The meta-goal for our course is to offer a meaningful, impactful, life-changing, and challenging intellectual experience that will enhance your ability to understand, think, create, and connect; and do so joyfully far beyond the course itself.

Puzzles:

Puzzles—of various types—will be used to provide minds-on practices of thinking, creativity, and tenacity that will be promoted throughout this course. The goal is that through your engagement with these puzzles, you will be able to transfer the thinking practices sharpened here to resolve the intellectual and day-to-day puzzles of your life.

At the beginning of each Monday evening class, you will submit written solutions to the three assigned puzzles. Each week, there will be one less-challenging, one medium-challenging, and one more-challenging puzzle. Puzzles announced as a "private puzzle" must be an individual effort—that is, you cannot discuss those puzzles with others (except with me). All other puzzles may be collaborative efforts with others from this class (but no one from outside our class). Collaborative work must be just that—a true collaboration; it cannot be a "you work on that puzzle, and I'll work on this puzzle" partner-ship. Moreover, you must think about and attempt any puzzle on your own *before* you engage in a collaboration with others. You will be graded not only on the correct-ness of the solution, but also on the clarity of the solution and your reflections on the effective thinking strategies applied to develop insights and generate epiphanies. *Thus, you are expected to submit edited, well-written final drafts.*

In addition to your Monday submissions, you will be asked to turn in a typed, two-paragraph reflection on the previous Friday's visitor. That report could in-clude a reaction, an insight, a connection with a previ-ous visitor, or an element of effective thinking the visi-tor employed.

Interim Status Reports. This course is unusual in many ways, including that it is more effective to create a clear, correct, and complete solution to a puzzle of which you are (justly) proud than to submit on time work that is incomplete and not reflective of your best effort. Thus, you are allowed to submit an "Interim Status Report" on any puzzle that you have not completed. That status report needs to clearly indicate the hours logged work-ing on that puzzle, all your false starts, previous attempts, and the effective thinking strategies applied together with what lessons were learned, as well as future strate-gies to move forward.

Practices of Effective Thinking:

Another unusual aspect of this course is that it is designed to directly impact your intellectual and creative efforts outside of our class. To this end, every Friday, you are to submit a copy of work (such as an assignment, paper, or reading) from another class in which you have intentionally and directly applied some element or exercise of effective thinking to that work. On that submitted copy, you need to clearly identify how your work was impacted by the practices of thinking from our course.

Papers:

There will be four essay assignments in this course and a final reflection/*Paideia* moment paper due during the final exam period. For each submission, you are to include the final typed version with the penultimate draft attached (showing your own handwritten edits).

Mindfulness:

This class focuses on habits of the mind to allow us to think, create, connect, live, and act more effectively; thus another unusual aspect of this course is an intentional mindfulness component. As a participant in this course, you agree to engage in a quiet mindfulness practice of your choosing for at least 10 wakeful minutes each day to help relax and recharge your mind as well as foster a positive mindset. Different options and suggestions for how to practice mindfulness will be offered.

Course Contract and Honor Code:

You will be asked to sign a contract of terms that you agree, on your honor, to follow in order to remain an active member of this class in good standing. For all puzzles, you are not allowed to consult with any Internet resources, texts, books, or others from outside this class.

Grading:

Puzzles	30%
Papers	25%
Practices of Effective Thinking	25%
Class participation	10%
Effective failure	10%
Total	100%

Effective Thinking through
Creative Puzzle-Solving
Contract

President Edward Burger
Southwestern University

I, _____, AGREE TO THE FOL-
LOWING TERMS:

1. I will experiment with new ways to think, create, and connect; I will understand more deeply, will strive to look at issues from many points of view, not give up because I am frustrated, learn from failed attempts, and develop a practice of creating questions.

2. I will come to class ready to actively engage in good spirits and in good humor, and I will try to apply lessons learned in this course to other aspects of my life.

3. I will try to engage in a peaceful, awake, mindfulness practice of my choosing for at least 10 minutes every day to recharge and renew my mind, as well as to foster a positive mindset.

4. All written work submitted will be carefully crafted so that my ideas are clearly expressed, and I will not submit a first draft of any writ-

ten work: All work will be the result of at least one revision and, if not typed, will be rewritten neatly.

5. All work submitted will be my very best work and best effort, and I will be proud of all my submissions.

6. For the entirety of each class meeting, I promise to be "off the grid"; that is, I will keep my personal electronic devices off and will not use any such devices during each class.

7. I will work independently on the "private puzzles," and for all puzzles, I will not consult with any Internet sources, texts, or others from outside this class.

8. I will not share the puzzles or their solutions with others from around campus in this and all future semesters.

9. I will abide by the honor code, and I will ask for clarification if I am not sure how it applies to any activity associated with this class.

Signed on this day, the _____ day of _____, 20___:

_____ . Witnessed by

_____

ACKNOWLEDGMENTS

This book would not have been possible without the meaningful and ongoing collaboration with my colleague and dear friend Michael Starbird. Mike has been an influential force for good in my professional journey as well as in much of my scholarly work and thinking. Mike continues to inspire me with his wisdom, creativity, joy of learning, and love of life; and I remain deeply grateful to be a beneficiary of his generosity of spirit.

Many current and former students and colleagues have influenced my work on this book, and I thank them all. In particular, I express my appreciation to all my students from *Effective Thinking through Creative Puzzle-Solving* courses at Southwestern University who provided feedback to improve the class as well as this book. Of special note, I thank Tristin Evans, Bryony McLaughlin, Aiden Steinle, and Jasper Stone, who not only made valuable suggestions, but also served as presidential interns in my office. Furthermore, I thank Kyle Brown, Elliott Foreman, Tanmai Korapala, and Kirhyn Stein for their insightful comments and ideas.

A number of colleagues and friends agreed to carefully read early manuscript versions of this text and provided important feedback and encouragement. In particular, I thank Michael Brewer, Florence Burger, John Chandler, Norma Gaines, Benjamin Holloway, William Powers, Paul Secord, and Fay Vincent. I am also grateful to all those who were class guest speakers and part of my *President's Thinking Symposium on Living, Learning, and Leading.* They not only offered meaningful lessons to my students on effective thinking, but also helped inspire aspects of this project. Those speakers are Herbert Allen Jr., Victor Bajomo, Ben Barnes, Carly Christopher, Clayton Christopher, Trammell Crow Jr., Paige Duggins, Michael Gesinski, Winell Herron, Philip Hopkins, Weston Hurt, Frank Krasovec, Red McCombs, Jessica Waldrop McCoy, Lynn Parr Mock, Presley Mock, John Oden, Ricky Raven, Valerie Renegar, Kendall Richards, Corbin Robertson Jr., Doug Rogers, Susan Slagle Rogers, Debika Sihi, Ken Snodgrass, and Tivy Whitlock.

I also thank Alisa Gaunder, my dean of faculty, for her encouragement, friendship, and support as I worked on this book; Lynn Parr Mock for taking it upon herself to entertain and uplift me during an intensive week-long writing session in Dallas; and Seamus and his parents for allowing me to share one of my favorite stories on the power of effective thinking.

Finally, it is a genuine delight to thank the Princeton University Press team of extraordinarily talented and creative professionals who are always open to new and creative ideas. As in my previous work, it has been a joy to collaborate with and learn from these wonderful individuals. Vickie Kearn, executive editor, has been a dear friend and an enthusiastic supporter of this project from the very beginning—mirrored chapter and all. Christie Henry, the Press's director, is an uplifting presence who is both connected and committed to the vision this work articulates. And from across the many departments of the Press, I wish to express my deep gratitude for the excellence of Bob Bettendorf, Lauren Bucca, Karen Carter, Lorraine Doneker, Sara Henning-Stout, Dimitri Karetnikov, Debra Liese, Stephanie Rojas, Susannah Shoemaker, Kathryn Stevens, Erin Suydam, and Kimberley Williams. I also wish to thank Karl Spurzem, who designed the beautiful dust jacket; Theresa Kornak, who skillfully copyedited my original manuscript; and Taylor Jones, Southwestern University alumnus from the great class of 1997, who artfully created the author photo.

INDEX

ABOUT THE AUTHOR

Dr. Edward Burger is president of Southwestern University as well as a professor of mathematics and an educational leader on thinking, innovation, and creativity. He has delivered over 700 addresses worldwide at venues including Berkeley, Harvard, Princeton, and Johns Hopkins as well as at the Smithsonian Institution, Microsoft Corporation, the World Bank, the International Monetary Fund, the U.S. Department of the Interior, the New York Public Library, and the National Academy of Sciences. He is the author or coauthor of over 70 research articles, books, and video series (starring in over 5,000 on-line videos watched by millions of viewers), including the bestseller *The 5 Elements of Effective Thinking*, with Michael Starbird, published by Princeton University Press and translated into over a dozen languages worldwide.

He has made over 50 television and radio appearances—at broadcast venues including ABC, NBC, Discovery, and NPR, and has been interviewed or quoted in the *New York Times*, the *Wall Street Journal*, *Newsweek*, the *Huffington Post*, the *Washington Times*, the *Austin American Statesman*, the *Houston Chronicle*, the *Chronicle of Higher Education*, and *Inside Higher Ed*. He has served as chair of the Board of the Associated Colleges of the South, a member of the Board of Directors of the Council of Independent Colleges as well as of the Governing Board of the Aspen Institute Wye Seminars in collaboration with the Association of American Colleges and Universities.

Burger has received more than 25 awards and honors for his work in education as well as in mathematics. In 2006, *Reader's Digest* listed him in their annual "100 Best of America" as America's Best Math Teacher. In 2010 he was named the winner of the Robert Foster Cherry Award for Great Teaching—the largest prize in higher education teaching across all disciplines in the English-speaking world. Also in 2010, he won a Telly Award for his appearance on an NBC-TV segment for the *Today Show* on the mathematics of the Winter Olympics. The *Huffington Post* named him one of their 2010 Game Changers, saluting, "100 innovators, visionaries, mavericks, and leaders who are reshaping their fields and changing the world." In 2012, Microsoft Worldwide Education selected him as one of their "Global Heroes in Education." In 2013, Burger was inducted as an inaugural fellow of the American Mathematical Society, and the following year he was elected to The Philosophical Society of Texas. He is now in the fifth season of his weekly program on thinking and higher education produced by NPR's Austin affiliate, KUT. The series, *Higher ED*, is available at kut.org/topic/higher-ed/ and on iTunes.